SAE EDGE™ RESEARCH REPORT

Unsettled Impacts of Integrating Automated Electric Vehicles into a Mobility-as-a-Service Ecosystem and Effects on Traditional Transportation and Ownership

Dr. Joachim Taiber
International Transportation Innovation Center (ITIC), USA
Dr. Raivo Sell
Tallinn University of Technology, Estonia

EDGE DEVELOPMENT TEAM

Dr. Robert Shorten, *Imperial College London*
Alexander Kraus, *TÜV SÜD*
Dan Baxter, *Stantec*
Ronald Grosse, *Bertrandt*
Dr. Reinhard Pfliegl, *A3PS*
Dr. Jim Martin, *Clemson University*
Mathew Yager, *IOTA Foundation*
Sean Kelley, *Mannik & Smith Group*
Hendrik Kramer, *Technical University of Munich*
Dr. Flavio Bonomi, *Nebbiolo Technologies and Lynx Software Technologies*

SAE INTERNATIONAL®
Warrendale, Pennsylvania, USA

About the Publisher

SAE International® is a global association of more than 128,000 engineers and related technical experts in the aerospace, automotive, and commercial-vehicle industries. Our core competencies are lifelong learning and voluntary consensus standards development. Visit sae.org

SAE EDGE™ Research Report Disclaimer

SAE EDGE™ Research Reports focus on topics that are dynamic, in which knowledge is incomplete, and which have yet to be standardized. They represent the collective wisdom of a group of experts and serve as a practical guide to the reader in understanding unsettled subject matter. They are not meant to provide a recommended practice or protocol. The experts engaged have contributed their own thoughts and points of view and these are not the positions of the institutions or businesses with which they are affiliated. A professional writer has collectivized their input; there is no one contributor's perspective being advanced, but rather that of a community of practitioners. SAE EDGE™ Research Reports are the property of SAE International and SAE alone is responsible for their content.

About This Publication

SAE EDGE™ Research Reports provide state-of-the-art and state-of-the-industry examinations of the most significant topics in mobility engineering. SAE EDGE™ contributors are experts from research, academia, and industry who have come together to explore and define the most critical advancements, challenges, and future direction in areas such as vehicle automation, unmanned aircraft, Internet of Things and connectivity, cybersecurity, advanced propulsion, and advanced manufacturing.

Related Resources

SAE MOBILUS® Automated & Connected Knowledge Hub
http://sae.org/Mobilus/Automated

SAE Team

Frank Menchaca, Chief Product Officer
Michael Thompson, Director, Standards, Information and Research Publications
Monica Nogueira, Acquisitions Director
Jill Leonard, Product Manager
William Kucinski, Managing Technical Editor

Printed in USA

ISSN 2640-3536
e-ISSN 2640-3544
ISBN 978-1-4686-0110-7

To purchase bulk quantities, please contact: SAE Customer Service

E-mail: CustomerService@sae.org
Phone: 877-606-7323 (*inside USA and Canada*)
724-776-4970 (*outside USA*)
Fax: 724-776-0790

Visit the SAE International Bookstore at books.sae.org

About the Editors

Born and raised in Germany, **Dr. Taiber** enjoyed his academic training at the Swiss Federal Institute of Technology in Zurich (ETHZ), Switzerland, where he first graduated with a master's degree in Mechanical Engineering in 1991 and then completed his PhD in Technical Sciences in 1996.

After his studies, Dr. Taiber started his professional career at a Swiss software start-up company in Zurich. He joined Bayerische Motoren Werke AG (BMW) Group in Munich in 1997 as in-house consultant in the vehicle development division where he worked on different aspects of functional integration and validation of vehicle systems, vehicle electronics platforms, and vehicle program management. He assisted the CEO as technology advisor and supported the CIO in a major reorganization of the Information Technology (IT) division which resulted in holding different leadership positions in the areas of IT strategy, program management, innovation management, and benchmarking.

In 2003, Dr. Taiber was engaged in the initial planning of an automotive research campus in Greenville, South Carolina, closely located to the BMW US manufacturing site that served as a public-private partnership model and helped shape strategic collaboration between BMW and Clemson University. The first facility created on the Clemson University International Center for Automotive Research (CU-ICAR) campus in 2005 was the BMW Information Technology Research Center (ITRC) - a dedicated building designed to collaborate with BMW IT partners on open innovation and strategic advanced technology projects including the aspects of the "networked vehicle" and the "next generation data center." Dr. Taiber led innovation activities at ITRC as director of the Information Technology Research Office for multiple years and worked closely with the BMW Tech Office in Silicon Valley, large technology companies, start-ups, and universities.

In 2010, Dr. Taiber joined Clemson University as research professor and member of the Automotive Engineering, Electrical and Computer Engineering, and Computer Science faculty. In 2012, he created the Advanced Vehicle-Infrastructure Technology Institute under the College of Engineering and Science, which he directed until the end of 2015. During his academic engagement, Dr. Taiber worked intensively on the development of vehicle electrification and charging infrastructure, connected vehicle systems, and intelligent transportation infrastructure.

In January 2016, Dr. Taiber was appointed Chief Technology Officer of the International Transportation Innovation Center, a nonprofit organization headquartered in Greenville, focused on developing and applying innovative methods of system validation for connected, automated, shared, and sustainable forms of mobility with industrial and academic partners. He also initiated a global consortium effort for the design and operation of smart mobility testbeds led by the Society of Automotive Engineers (SAE International) - the International Alliance for Mobility Testing and Standardization (IAMTS).

Through his private consulting practice, Dr. Taiber is involved in the development, implementation, and operation of open and closed mobility testbeds both at a national and international scale. Additionally, through a private venture practice, he engages in the development of new mobility-related businesses.

He continues to serve Clemson University as adjunct professor at CU-ICAR. Dr. Taiber still acts as a smart mobility advisor of the IOTA Foundation (blockchain and distributed ledger technologies) and is active in transportation-related standards development efforts both at SAE and IEEE. He is also member of the advisory board of the Peace Innovation Foundation which works closely with the United Nations.

Raivo Sell received his PhD in Product Development from Tallinn University of Technology in 2007. His research interests cover early design issues on mobile robotics and self-driving vehicles. He currently runs the Autonomous Vehicles Laboratory-Competence Center at TalTech as a research group leader.

Raivo Sell has been a visiting researcher in ETH Zürich, Aalto University, and, most recently, at Florida Polytechnic University in the USA and recognized as a Chartered Engineer and International Engineering Educator: ING-PAED IGIP. He is an author of several books on the Internet of Things and practical robotics, and has been an educator of robotics teachers in vocational and higher education.

contents

Unsettled Impacts of Integrating Automated Electric Vehicles into a Mobility-as-a-Service Ecosystem and Effects on Traditional Transportation and Ownership

DR. JOACHIM TAIBER

International Transportation Innovation Center (ITIC), USA

DR. RAIVO SELL

Tallinn University of Technology, Estonia

Abstract

The current business model of the automotive industry is based on individual car ownership, yet new ridesharing companies such as Uber and Lyft are well capitalized to invest in large, commercially operated, on-demand mobility service vehicle fleets. Car manufacturers like Tesla want to incorporate personal car owners into part-time fleet operation by utilizing the company's fleet service. These robotaxi fleets can be operated profitably when the technology works in a reliable manner and regulators allow driverless operation.

Although Mobility-as-a-Service (MaaS) models of private and commercial vehicle fleets can complement public transportation models, they may contribute to lower public transportation ridership and thus higher subsidies per ride. This can lead to inefficiencies in the utilization of existing public transportation infrastructure. MaaS platforms can also cause a reduced reliance on parking infrastructure (e.g., street parking lanes and parking garages) which can contribute to an improvement in overall traffic flow, and a reduction in capital investment for commercial and residential real-estate development. Urban planning can be better centered around the true mobility needs of the citizens without sacrificing valuable space for vehicles that are inactive most of the time. A key challenge is the transition phase where traditional forms of car ownership and transportation coexist with new forms of mobility services (before true MaaS platforms are fully optimized). Another fundamental issue is determining the true cost of car ownership and establishing ideal cost structures for on-demand mobility services in order to replace the need of owning a car. This needs to occur without compromising subsidized public transportation which provides affordable transportation for low-income groups.

NOTE: SAE EDGE™ Research Reports are intended to identify and illuminate key issues in emerging, but still unsettled, technologies of interest to the mobility industry. The goal of SAE EDGE™ Research Reports is to stimulate discussion and work in the hope of promoting and speeding resolution of identified issues. SAE EDGE™ Research Reports are not intended to resolve the issues they identify or close any topic to further scrutiny.

Edge Development Team

Dr. Robert Shorten, *Imperial College London*
Alexander Kraus, *TÜV SÜD*
Dan Baxter, *Stantec*
Ronald Grosse, *Bertrandt*
Dr. Reinhard Pfliegl, *A3PS*
Dr. Jim Martin, *Clemson University*
Mathew Yager, *IOTA Foundation*
Sean Kelley, *Mannik & Smith Group*
Hendrik Kramer, *Technical University of Munich*
Dr. Flavio Bonomi, *Nebbiolo Technologies*

ISSN 2640-3536

Introduction

State of the Industry

While the automotive industry continues to grow in terms of global production, the United States (US) and European markets are becoming saturated. The infrastructure in these markets are at capacity regarding traffic flow in many urban regions, and searching for parking space is taking valuable time away from drivers.

However, the American Automobile Association (AAA) Foundation for Traffic Safety recently indicated that replacing a personal vehicle for a person living in a city with an average of about 10,000 miles driven with ride-hailing services would roughly double the cost of transportation [1]. The cost per mile for a personal vehicle will average around $0.93/mile (including parking fees) whereas the cost of ride hailing (plus use of car rental for longer trips) will average $1.86/mile.

Owning a car is, on average, still less expensive than ride hailing, but the de facto utilization of a personal vehicle is low. Generally, cars are parked about 95% of the time [2]. As the average occupancy rate of a passenger vehicle in the USA is less than two people, it is fair to say that privately used vehicles are an underutilized asset class, which contribute to traffic congestion. As an alternative to the private car model, the rideshare model could replace personal cars through fleet cars that are in operation all the time, and which can pool ride to increase the occupancy rate of the vehicle. Another option is to share a private car with others (carpooling) or completely rent it out to third parties to increase utilization (Figure 1).

According to a recent McKinsey Report, the miles traveled through ridesharing reached two billion per year with an annual growth rate of 150% between 2013 and 2018 [3]. However, a key problem is that ridesharing is not replacing other forms of transportation to 100%, about 50% are new miles which contribute to additional traffic (Figure 2).

Another important metric is the average trip length of a shared ride; the average rideshare trip distance in 2015 was around 6 miles [4]. If we compare this to the findings of the 2017 National Household Travel Survey from the US Department of Transportation Federal Highway Administration, the average trip length with a personal vehicle across all purposes was about 10 miles [5]. It is important to note that half of all trips are 3 miles or less and are driven by 72% drivers and even trips of less than 1 mile are driven by 60% [6] (Figure 3).

Although the ridesharing model has been adopted by consumers and the market is growing at a global scale [7], the behavior of using a personal vehicle is not changing in the USA. This can be traced back to the fact that owning a vehicle is still affordable for the majority of the population

FIGURE 1. **Average vehicle occupancy by vehicle type according to the US Department of Energy's Office of Energy Efficiency and Renewable Energy.**

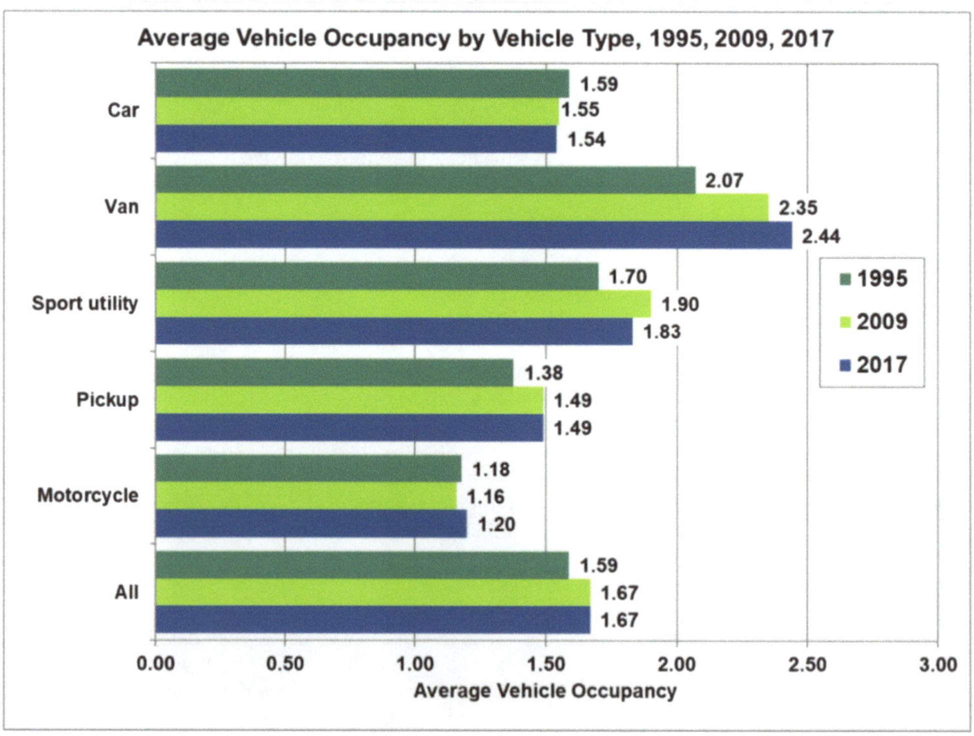

FIGURE 2. **E-hailing growth sources. Exhibit from [3].**

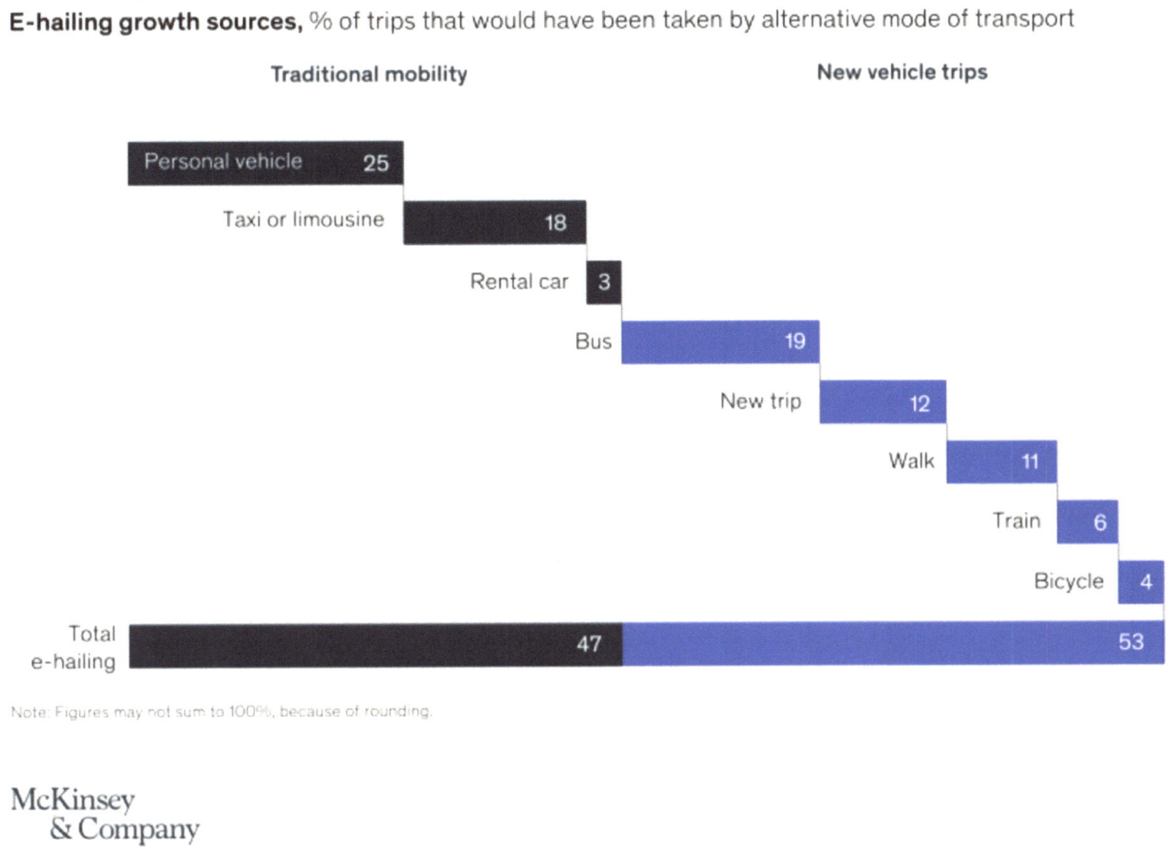

FIGURE 3. **Average US Uber trip length increased by 12% in 2015.**

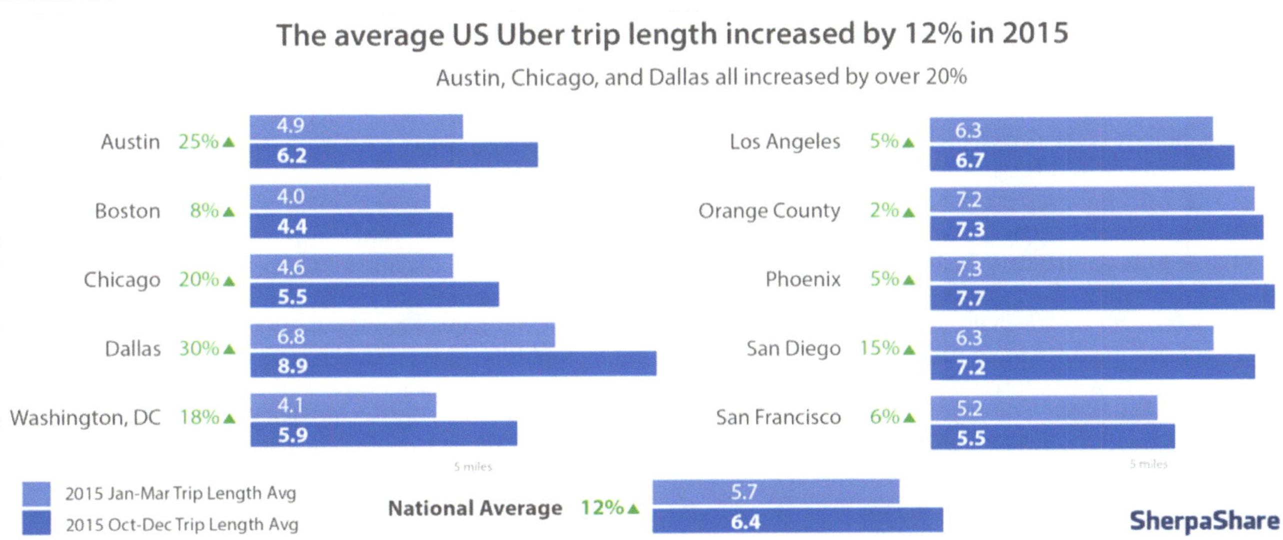

and it provides a feeling of privacy and independence which is typically compromised through ridesharing.

A recent AAA study confirms that teenagers are waiting longer before obtaining a driver's license. The study indicates that this behavior is influenced by the availability of ridesharing services [8]. At the other side of the age spectrum, ridesharing can be a useful transportation alternative for older people as their ability to drive a car deteriorates. However, a recent AARP study found that 94% of surveyed Americans above the age 50 have heard of ridesharing services, but only 29% have ever used ridesharing services, citing safety and privacy concerns [9].

In addition to consumer habits, the influence of new automotive technologies such as self-driving technologies and vehicle electrification are impacting the development of ride-hailing business models. The development of self-driving vehicles has been fueled by significant advances in artificial intelligence (AI) technologies, sensor technologies, and available computing power [10]. A key problem is the endless variety of situations between dynamic and static objects that can occur in the real world of traffic and the limitation of processing power for high-resolution signals. These challenges make it difficult to produce a self-driving vehicle that performs better than the average human driver on the road. Although the car and the technology industry have invested billions of dollars into the development of self-driving technologies, currently no full general deployment of self-driving vehicles exists in operation except pilot programs in well-defined geographical areas.

What could help with an accelerated role out of self-driving vehicle technology is teleoperation where the vehicle can be controlled remotely if needed via a low-latency network connection [11].

One of the biggest economic hurdles for the introduction of self-driving technology for ride-hailing operators is the use of safety drivers. As ride-hailing operators are considering using vehicles without steering wheels and pedals, teleoperation can ensure safety at all times, utilizing human operators in the backend.

Another challenge for the implementation of self-driving vehicle technology is the need for a regulatory environment both on the vehicle and the infrastructure side, including testing standards. But even if rules and regulations for self-driving vehicles are in place, the immediate future will see a mix of human-driven vehicles and machine-driven vehicles on the road with an increasing portion of machine-driven vehicles. This means that vehicle behavior and traffic partners will not be fully predictable, which increases the risk of misinterpretation both on the human as well as on the machine side.

The automotive industry is currently preparing for a true mass-adoption of electrified vehicles. This is driven by dropping battery cost per kilowatt hour (kWh) and the effects of economy of scale on electric motors and powertrain control systems. Whereas in 2015, more than 50% of the total cost of a midsize electric vehicle was related to the battery, for 2019 this number is 33%, and by 2025 it is estimated to be around 20%. Additionally, the cost parity between electric cars and internal-combustion engine powered vehicles is predicted to be achieved by 2022 by progressive analysts [12].

Traditional car manufacturers will be presented with the challenge of retooling their manufacturing plants and adapting their vehicle platforms to support all-electric vehicles without compromising available space for passengers or luggage. At the same time, the public and workplace charging infrastructure needs to be developed to support a large fleet of electrified vehicles on the road. In order to provide the energy load for the infrastructure, electric utilities need to provide more than 700 terawatt hours by 2030 [13]. Fast-charging technologies are being rolled out, which would allow for a 200- to 300-mile charge in 15 minutes (using up to 500 kW of direct current charging) [14] (Figure 4).

In 2018, the University of Michigan found that the average cost to operate an electrified vehicle in the USA is about $485 per year, whereas the average operating cost for a gasoline-powered vehicle is $1,117 [15]. Additionally, the maintenance cost of electrified vehicles are lower, with less moving parts, less fluids, no spark plugs, and less brake wear through regenerative braking [16].

However, ownership costs for vehicles with internal combustion engines are lower than electrified vehicles for high-mileage fleet applications. Also, the market shows that ride-hailing companies are lagging in introducing electric vehicles into service. In 2018, less than 1% of the vehicles on the Uber and Lyft platforms were electric [17] (Figure 5).

One of the biggest challenges for the automotive industry is the change of its business model from "selling a vehicle" toward "selling a mobility service." Selling a car to a private customer is very different from selling a car to a commercial fleet operator. Also selling a mobility service via a digital channel and delivering end-to-end service on a mobile phone is very different from selling a traditional bus ticket. The key question in this transition of the business model is the impact on competitive forces. Traditionally, automotive original equipment manufacturers (OEMs) compete on aspects such as vehicle design, vehicle performance, and retail cost. Self-driving features and electrified driving capabilities have been first introduced to the market as expensive extras, which are limited to premium vehicles. Through economy of scale, those capabilities become more affordable and can be applied across all vehicle categories. Ultimately, these capabilities will be driven by regulation, which OEMs must comply with. However, OEMs are starting to collaborate on aspects such as developing high-definition maps, manufacturing battery cells, and sharing AI, mobility service, and electric vehicle charging platforms to reduce development cost and to accelerate mass adoption.

There is also the potential to combine public transportation with multimodal forms of transportation in innovative ways. For example, could ride hailing be combined with mass transit to achieve a more economic operation both on the client and the operator side? Could this also include the use of bicycles or scooters?

Ultimately, a true MaaS-based transportation model requires both change of user behavior and adaption of the infrastructure.

FIGURE 4. **Public and workplace charging infrastructure in place in 2017 as percentage of infrastructure needed by 2025 by metropolitan area.**

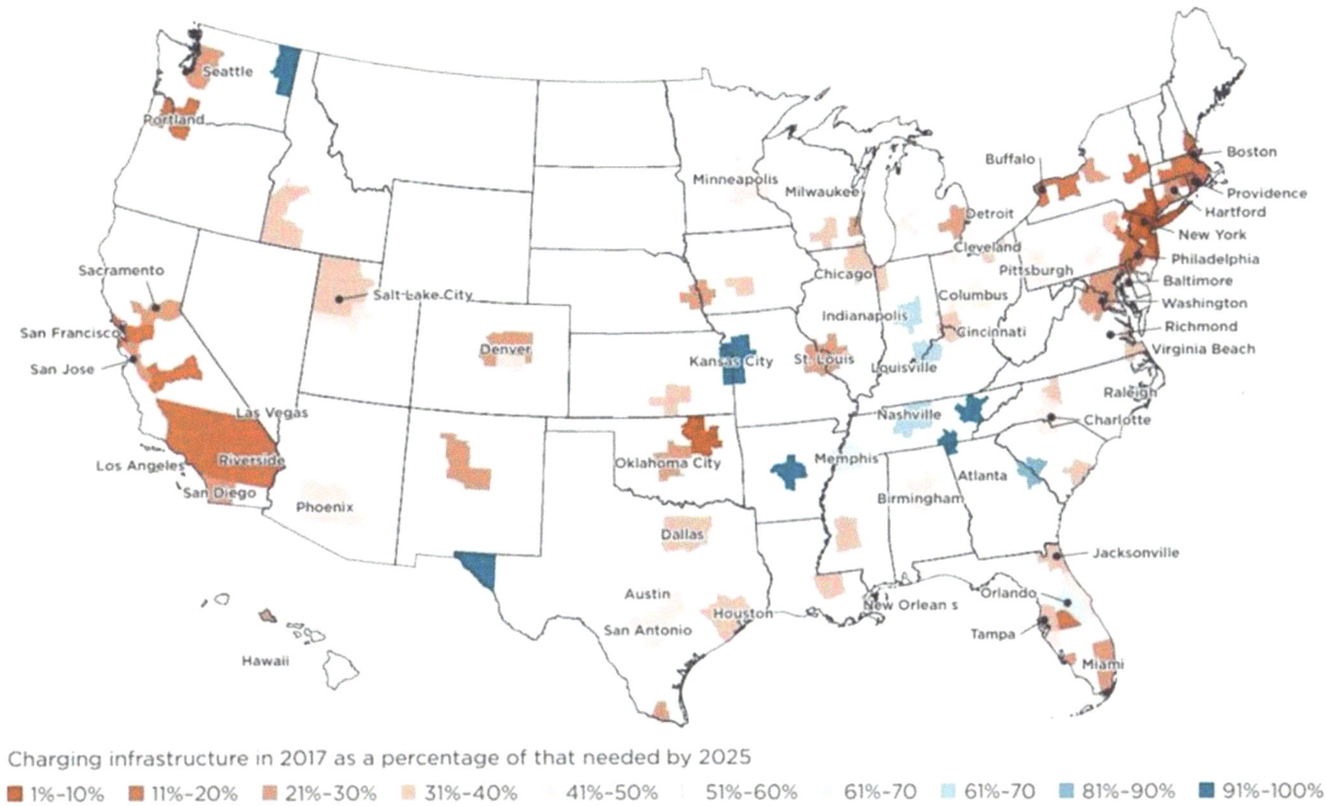

Unsettled Issues Concerning Integrating Automated Electric Vehicles into a MaaS Ecosystem and Effects on Traditional Transportation and Ownership

- What is the true cost of automated driving and how quickly is it being adopted?
- AI and regulations are still lagging regarding fully automated vehicle (AV) deployment.
- Electric-vehicle mass adoption has a significant impact on the MaaS ecosystem.
- In which areas of self-driving technology should OEMs compete and collaborate?

True Cost of Automated Driving and Speed of Adoption

What is largely unsettled is the timeline and the impact of electrified robotaxi fleets regarding MaaS platforms. In particular, technology robustness, infrastructure readiness, and consumer acceptance - and cost of service and service availability - must be examined. These aspects are highly interrelated.

A recent study from Capgemini reveals that consumer acceptance is growing with a majority preferring to ride in a self-driving car versus a normal one within five years [19]. Consumers see the biggest benefit of self-driving cars in saving time. It was also found that close to half of early adopters are willing to pay a premium of more than 20% of their current travel budget to use self-driving cars. The survey also showed that traditional OEMs receive more consumer trust regarding self-driving car manufacturing than start-ups (Figure 6).

There is no doubt that the investment community played a major role in enabling start-ups to work on innovations

FIGURE 5. **Cost of ownership. Exhibit from [18].**

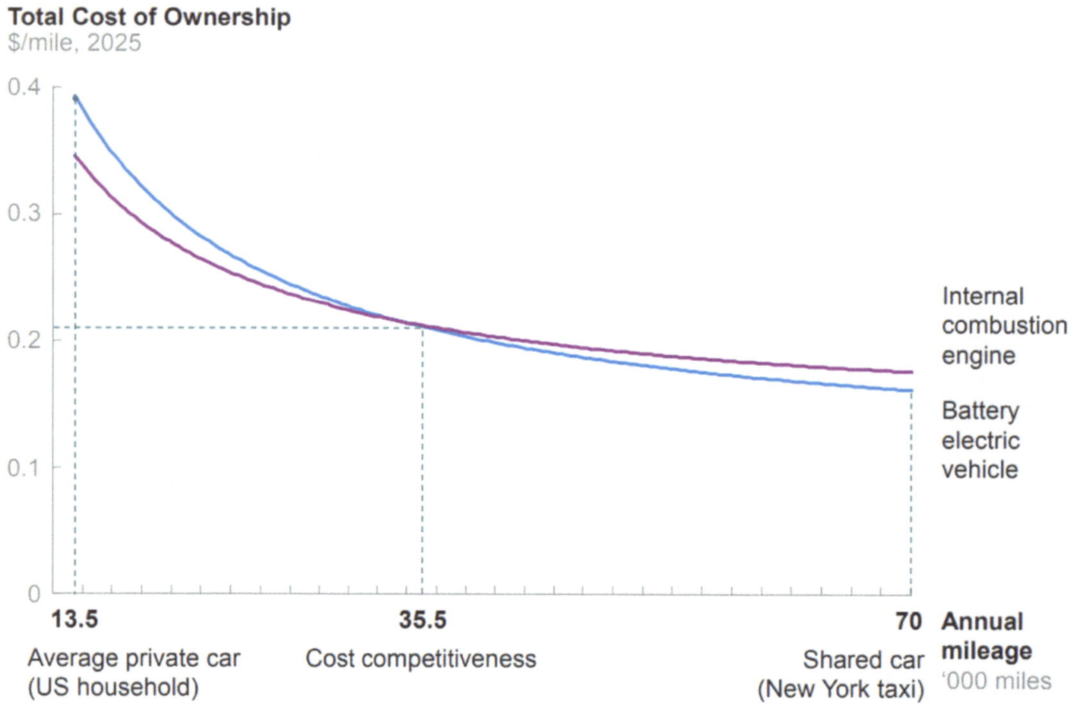

required for self-driving vehicles which in return need to work with automotive suppliers and OEMs to validate their solutions. Investments in the self-driving car segment surged from $6 billion in 2015 to more than $60 billion in 2018 [20]. However, important industry leaders such as the CEOs of Waymo and Ford are indicating in recent statements that the mass deployment of self-driving vehicles will take longer than expected due to the complexity of the matter. The current dilemma for the automotive industry is that market projections about miles travelled by robotaxis can only materialize if the robustness of the systems (in terms of safety and security) can be guaranteed.

There is also a huge discrepancy in market projections about the cost per mile of operating a robotaxi. Telsa's Elon Musk compared the average cost per mile of ridesharing of $2.00 to $3.00 with the expected cost per mile of a Tesla robotaxi fleet of being less than $0.18. What is important to understand is that Tesla wants to operate private-owned SAE Level 4-capable Tesla vehicles as robotaxis so they do not have to shoulder the capital cost [21]. This model also allows Tesla owners to generate revenue from their car ownership.

On the other hand, a recent research study from the Massachusetts Institute of Technology (MIT) estimates that the cost per mile to operate autonomous vehicles with single occupants ranges between $1.58 to $6.01 per mile - significantly higher than the average cost of personal car ownership (approximately $0.59 per mile according to US Bureau of Transportation Statistics) [22]. To improve the economics, robotaxi utilization rates would have to be substantially increased, which would trigger trip discounts for passengers based on length of trip and number of occupants. The biggest cost drivers are autonomous-vehicle licensing and safety oversight. Vehicle cleaning also needs to be considered. An overview about the cost breakdown of the autonomous vehicle operating cost in the MIT study can be seen in Figure 7. Through standardization of safety and security assessment procedures, AV licensing costs can be reduced. As indicated earlier, teleoperation could be a way to get the physical safety drivers out of the vehicle and to reduce the cost of fleet operation (Figure 8).

Another aspect to consider in the context of robotaxi is the cost of the vehicle itself. Although Uber and Volvo Cars recently presented a production ready AV [24] and Uber

FIGURE 6. **How consumers rate their overall level of trust [19].**

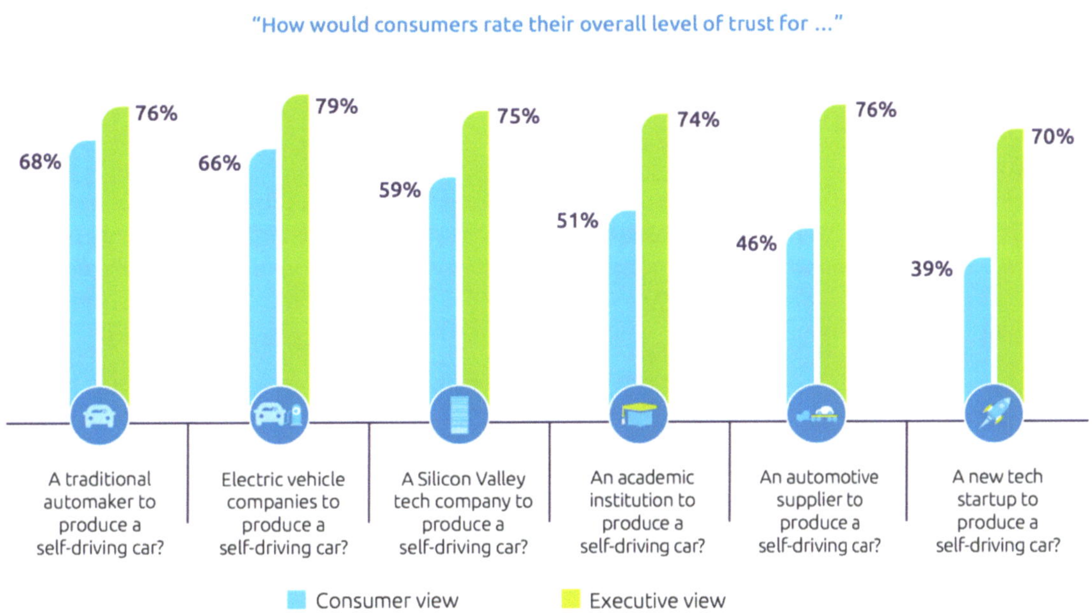

Reproduced with permission from Capgemini Research Institute

FIGURE 7. **Robotaxi deployment phases. Exhibit from [3].**

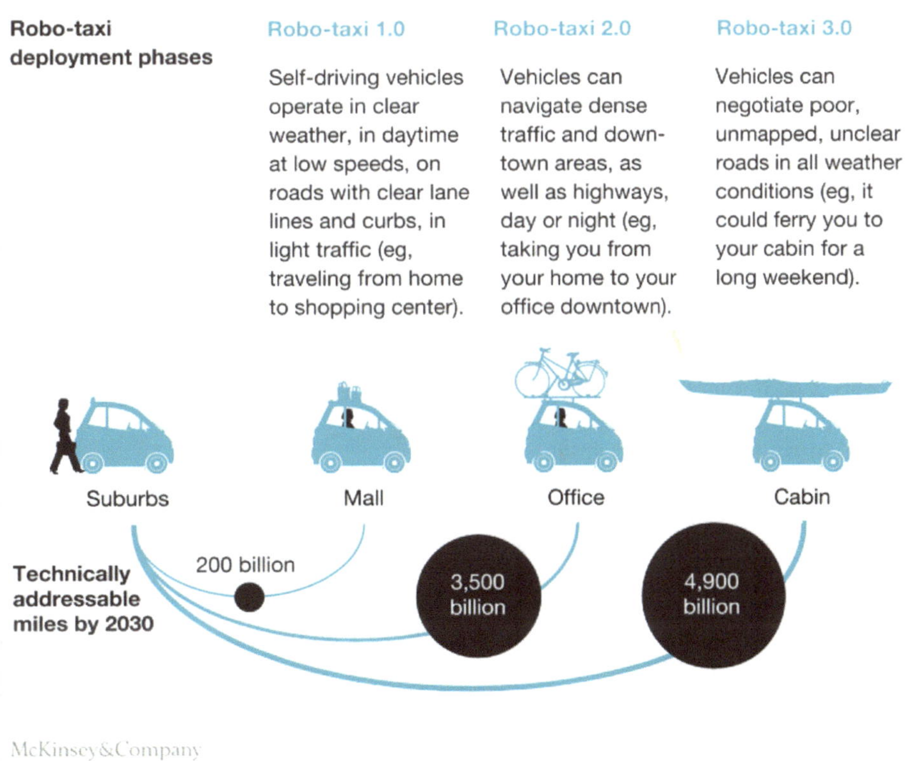

FIGURE 8. **Autonomous taxi fare estimation.**

	Total Cost	Cost per Mile
Vehicle Financing	$15,000	$0.04[i]
Licensing	$250,000	$0.16[j]
Insurance	$9,600 p.a.	$0.11[k]
Maintenance	$7,110 p.a.	$0.08[l]
Cleaning	$4,520 p.a.	$0.05[m]
Fuel	$6,305 p.a.	$0.07[n]
Safety Oversight	$211,662 p.a.	$0.05 - $2.35[o]
Profit	$24,363 p.a.	$0.27[p]
Fare	n.a.	$1.58 - $6.01[q]

MIT

Reproduced with permission from Ref. [23]

ordered more than 20,000 vehicles from Volvo to be used in their robotaxi fleet [25], the current cost of L4 vehicles can easily add more than $100,000 to the base price of a vehicle. Even with economies of scale, there will be a cost premium [26]. As the vehicles are planned to be in operation all the time and thus highly utilized, the time for amortization might be reasonable but this will be influenced by customer acceptance and achievable price per mile.

Recommendations

The largest cost factor in robotaxi fleets are safety drivers. Removing them is essential to implement teleoperation services. However, those services require a high-performance network and cloud-control infrastructure which might not always be available, especially in rural areas.

To gain customer acceptance, the automated driving service needs to be reliable and safe. This can be demonstrated best via open testbed implementations which allow users to experience the service under real-world conditions. This can start with low-speed automated shuttles that use dedicated lanes to engage users in the self-driving experience while replacing short trips normally taken in personal cars. Then, as OEMs learn to manage complexity, combine low-speed automated shuttles with higher speed robotaxi fleets through the adoption of MaaS platforms.

Other methods could involve forming specific urban zones which do not provide parking for personal cars and which can only be served through robotaxi solutions. Here, robotaxi fleets could be integrated with public transportation solutions (e.g., last-mile transport).

Additionally, two other strategies would be used:

- Adapt road infrastructure and road use regulations for robotaxis to enable faster commute times (e.g., dedicated lanes).
- Validate business models for automated shuttles to operate at cost parity with conventional shuttles.

Lagging AI and Regulations

The interaction between vehicle and infrastructure can be explained best using the example of low-speed, self-driving electrified shuttles in smart city environments. For these first- and last-mile mobility services, the use of slow-driving autonomous shuttles can complement both public transportation and personal vehicles, and they can also help to move parking areas into less crowded urban zones (Figure 9).

However, those vehicles will require dedicated lanes which need to be integrated into the existing road infrastructure (like bus lanes). Cities need to be willing to make the investment to adapt the road infrastructure and to

FIGURE 9. **ISEAUTO is the first self-driving vehicle developed in Estonia.**

Reproduced with permission from TalTech

reconsider existing parking regulations. Furthermore, the availability of Vehicle-to-Everything (V2X) communication infrastructure is a key enabler for the safe operation of autonomous shuttles. Data that is provided by the city (e.g., digital road rules), connected vehicles, and/or pedestrians can be utilized by the vehicles, and data generated by the vehicle sensors can be transmitted to the cloud and shared with other traffic participants if applicable, and in compliance with safety and privacy regulations. Careful consideration needs to be given to the cooperation of multiple AVs and the resilient design of the underlying V2X network which needs to be able to handle impairments such as latency and loss of messages to complete network outages. Another important aspect is to keep transaction data records (e.g., billing, maintenance, tracking) safe from manipulation which could be organized through blockchain-based methods where all transactions are authenticated.

As autonomous shuttles go through the perception-planning-control cycle, they need to identify existing and potential driving scenarios using sensors and AI algorithms. Even in geofenced areas with predetermined routes, the complexity of maintaining a safety level that is at the level of a human driver or better is enormous. For mass deployment, all potential scenarios need to be known as well as standardized. The vehicle needs to be tested against those scenarios both in simulation as well as in the real world (as long as this can be done in a safe manner), and the reaction patterns need to satisfy well-defined conditions.

Both a digital model of the vehicle and the infrastructure it operates in needs to exist and needs to be current in the data that is used to describe it. The mapping of the digital world into the physical world is challenging as the functional model of the digital vehicle does not always fit exactly to the functional behavior of the real vehicle. The biggest problem is misinterpretation or inability to interpret a specific situation the vehicle is confronted with. This can lead to accidents or end of operation. Teleoperation provides the ability to monitor the vehicle and prevent those situations, allowing a human operator to take over if needed.

At high speeds, more information needs to be processed to evaluate the potential dynamic interaction of objects, and this requires the use of high-performance computing processors. Additionally, the system is more robust when the physical environment and potential interaction patterns are modeled in high quality. Thus, low-speed autonomous shuttles in well-controlled environments are a good starting point to validate a robotaxi fleet operation. For proper implementation of a low-speed autonomous shuttle service, the road infrastructure needs to be properly adapted by analyzing (and, if possible, eliminating) potential safety risks. Also, a reliable V2X infrastructure needs to be built to allow for teleoperation, to map vehicle capabilities, and to identify and process driving scenarios. These scenarios need to be recorded under the requirements of the deployment environment which needs to be proven in an independent safety assessment.

A major problem is the absence of standardization regarding test procedures, scenarios, and test infrastructure. These are currently being addressed by consortia such as the Automated Vehicle Safety Consortium (AVSC) and the International Alliance for Mobility Testing & Standardization (IAMTS). Standards need to be defined for all vehicle and infrastructure categories. Both state- and federal-level regulators are being asked to create the legal framework which allows OEMs, suppliers, infrastructure service providers, and operators to have a predictable environment for their investments. Infrastructure investments are typically triggered when standards are being defined. On the international level, the harmonization of standards is important to increase economy of scale of key components.

The introduction of self-driving vehicle technology to the market is primarily being challenged by reliability and safety concerns as well as the lack of regulatory framework to support mass deployment.

Recommendations

Standardization of driving scenarios across the connected and autonomous vehicle ecosystem is important to establish a common language of understanding how vehicle systems should react to specific driving situations.

Intelligent road systems need to be classified for which driving scenarios they can support. This would allow vehicle manufacturers and infrastructure service providers to define standardized test procedures to validate system behavior.

Cities should consider establishing dedicated lanes for Level 3- and Level 4-enabled vehicles to make driving behavior more predictable and reduce complexity. Specific road usage rules for the interaction between highly AVs could allow for platooning at constant speed and checking vehicles in and out of platooning lanes.

In order to test and validate highly automated systems, both closed and open testbeds are needed. Those testbeds must be compatible through the support of standardized scenarios and full support of teleoperation through adequate V2X infrastructure.

To ensure compliance with intelligent road usage rules, highly AVs need to be connected to cloud-based traffic control systems. This could enable the fee-per-mile model for road usage and eliminate the risk of running out of fuel or battery power and for emergency response.

Furthermore, cybersecurity risks could be better detected and the potential impact for traffic participants be reduced. It is recommended to consider the use of blockchain technologies for the implementation of intelligent automated fleet operation and secure storage of transactional data.

Electric Vehicle Mass Adoption Has a Significant Impact on the MaaS Ecosystem

A 2018 survey from AAA shows that 20% of Americans are likely to buy an electric car, and consumer interest is continuously growing [27]. High reliability ratings, increases in range, and the availability of an electric-charging infrastructure will continue to strengthen consumer confidence. A recent Deloitte study also confirms that electric vehicles are continuing to penetrate the global market at an exponential rate [28].

In the USA, low fuel prices are slowing the demand for electric vehicles. There are also substantial differences in consumer concerns related to electric vehicles across key markets.

Important events in the push toward electrification include "Dieselgate" and the emergence of Tesla Model 3 as a high-demand, mass-produced electric vehicle. Latest research from the UBS Group indicates that the cost of ownership of diesel cars will be higher than battery-electric vehicles from the early 2020s onwards. And, as production capacities for battery cells and battery packs are rapidly increasing worldwide, the cost curves for battery packs are declining (Figures 10 and 11).

FIGURE 10. **Cost of ownership of diesel cars are likely to turn into a disadvantage (annual costs in EUR).**

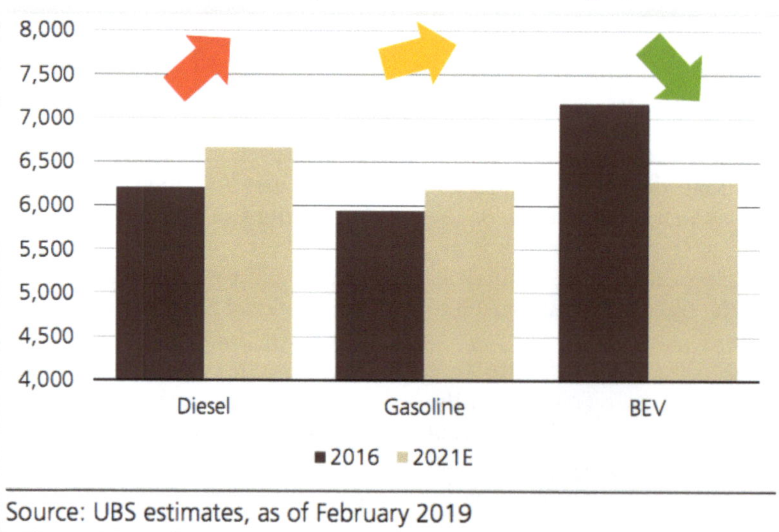

FIGURE 11. **Cost of ownership per kWh.**

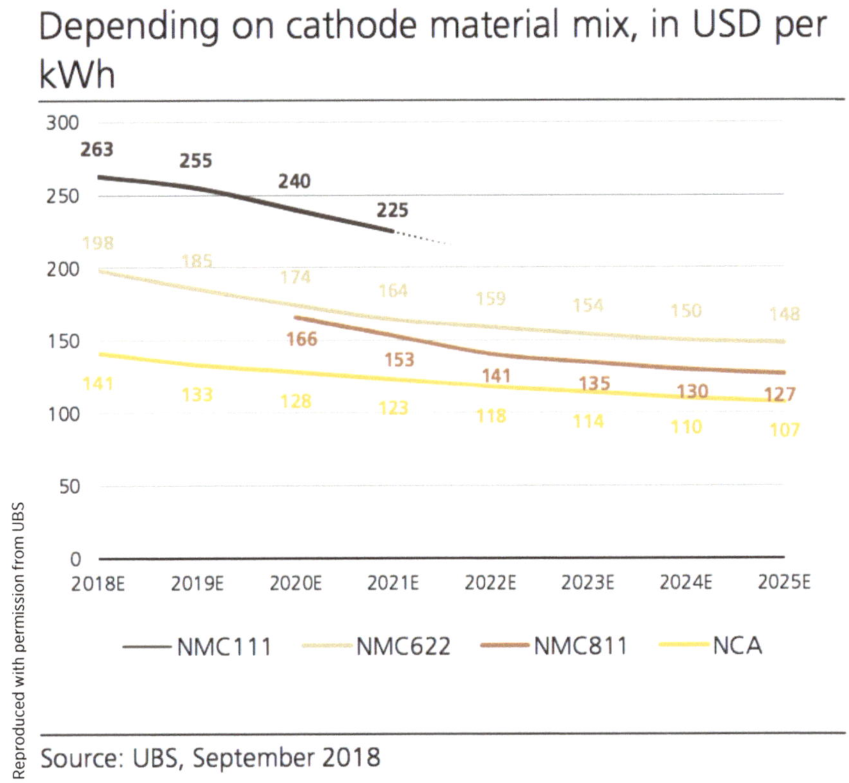

It is likely that future fleet vehicles used in MaaS platforms will be predominantly electric, which means that it is very important that an adequate charging infrastructure is in place. One of the challenges for electrified robotaxis is that they need to be on the road with minimum service and maintenance interruptions, so the charging process needs to be automated and time optimized. This could also require the use of technologies such as ultrafast, high-power charging and wireless charging, as well as a proper service infrastructure for large fleets of electrified vehicles.

As awareness of the environmental impact of driving grows, it may be important for consumers to know whether a fleet vehicle is electric and whether the energy it uses was produced by renewable sources.

Recommendations

Fleet operators using MaaS platforms need to plan the large-scale adoption of electrified vehicles. This could be driven both by cost considerations as well as consumer demand.

Electric vehicle infrastructure development needs to consider not only the use of private electric vehicles but also commercial electric vehicles.

Training of service technicians with skills in electrified vehicles and infrastructure needs to be addressed.

Standards development for fast, high-power, and wireless charging needs to be enforced.

Competitive and Collaborative Forces for Automotive OEMs Regarding Self-driving Technology

A major challenge for automotive OEMs is to decide where to compete and where to collaborate in the field of self-driving technology. Even companies that invested billions of development dollars into the technology such as Waymo, Uber, or GM have not been able yet to start mass deployment. Instead, collaborations and alliance have formed to share the cost of development and to define standards [29].

Once the regulatory and technical hurdles of robotaxi fleet deployment are overcome, there is no doubt that MaaS platforms will become the primary instruments for the interaction between customers and operators. We can already see today that dominant players in the ride-hailing market such as Uber are horizontally integrating to offer a wide variety of mobility services from food delivery, passenger transport, ground transport, and air transport, so it is very likely that there will be competition between the platforms similar to the competition between operating system platforms for mobile phones.

What we have seen is a competition between traditional automotive OEMs to develop self-driving technology and technology companies integrating self-driving technology into existing vehicles. In rare cases, technology companies are developing a self-driving car platform from scratch. Automotive OEMs, Tier 1 suppliers, and mobility service providers such as Uber and Lyft are hiring software and hardware experts to develop critical knowledge in self-driving technology. At the end of the day, the role of the OEM is to supply AVs to mobility service providers which use MaaS platforms. Efforts of OEMs to become global mobility service providers have not been very successful so far, an example is the Reach Now service [30]. On the other hand, efforts of technology companies and mobility service providers to deploy electrified robotaxi fleets require intensive collaboration with OEMs. This is seen in the example of Uber and Volvo Cars.

A major challenge is the transition from public mobility service providers, which are typically subsidized to offer affordable transportation, to private mobility service providers. This is particularly delicate because private mobility service providers typically use public road infrastructure. Instead of competition, a better concept is the collaboration of public and private mobility service providers (e.g., to support first-mile/last-mile transportation).

Recommendations

Connected and autonomous vehicle ecosystem providers need to collaborate on standards that contribute to cost reduction in development and testing of self-driving technology in order to pave the way for mass adoption.

It is better for OEMs to focus on supplying self-driving vehicles to mobility service providers with global reach that operate the MaaS platforms.

Public and private mobility service providers should collaborate to offer customers a complementary portfolio of transportation services that are cost effective and sustainable.

Summary/Conclusions

The use of a personal car is more convenient than using public transportation, and it is still less expensive than using a ride-hailing service. However, as the majority of Americans are driving to work by car, traffic congestion is inevitable.

The topic of whether robotaxis can be operated at a substantial lower cost than a personal vehicle is largely unsettled and is being challenged by a recent MIT study. However, an opportunity exists in the market for the use of low-speed, electric-automated shuttles for trips shorter than two miles. While this may require the use of dedicated road infrastructure, it will provide an entry point for consumers to test the technology before multiuser, high-speed robotaxis are offered through a MaaS platform. This only works if AV services can be offered at a competitive cost and with a very high level of safety and security.

We also have the opportunity to make large corrections in the way the current transportation infrastructure is designed which reserves a large amount of space for parking in, especially in urban areas. It is likely that robotaxi fleets offered by private operators (and then later by public transportation operators) will be deployed and reduce the need for parking infrastructure. However, it is important to optimally integrate public transportation services with mobility service operators via suitable MaaS platforms and public infrastructure changes.

The mobility service of the future will be connected, automated, electric, and shared. Ultimately, there will be less need to own their personal vehicle. It will be a choice.

SAE EDGE™ Research Reports

SAE EDGE™ Research Reports, like the present report on "Unsettled Impacts of Integrating Automated Electric Vehicles into a Mobility-as-a-Service Ecosystem and Effects on Traditional Transportation and Ownership," are intended to push further out into still unsettled areas of technology of interest to the mobility industry. SAE launches these reports before attempting to form a joint working group, let alone a Cooperative Research Program or a Standards Committee.

SAE EDGE™ reports are intended to be quick, concise overviews of major unsettled areas where vital new technologies are emerging. By definition, an unsettled area is characterized more by confusion and controversy than established order. Early practitioners must confront an absence of agreement. Their challenge is often not to seize the high ground but to find common ground. These scouting reports from the frontiers of investigation are intended merely to begin the process of sorting through critical issues, contributing to a better understanding of key problems, and providing helpful suggestions about possible next steps and avenues of investigation.

SAE EDGE™ Research Reports, therefore, are fundamentally distinct from the more formal working groups approach and far removed from the more mature Cooperative Research Program and standard's development process. They begin by convening an informal meeting of subject matter experts working in an unsettled area of technology from industry, government, and the research community. A moderated but collegial discussion ensues regarding their shared interest. This conversation attempts to identify critical issues hindering a clear understanding of important topics, preventing a common agreement on potential solutions, or slowing progress in key areas of research.

An SAE EDGE™ Research Report is then drafted and circulated for comment to meeting attendees and other interested parties. After review, the final result of this process is a report like the current examination of integrating automated electric vehicles into a MaaS ecosystem.

Next Steps for Integrating Automated Electric Vehicles into a MaaS Ecosystem

This publication should be considered only as a first step toward clarifying the issues around automated electric vehicles and MaaS integration. The intention behind this and other SAE EDGE™ Research Reports is to start a dialogue among interested parties on important industry-wide topics that require further attention. The expectation is that these explorations of unsettled areas of technology will lead to the formation of working groups and, ultimately, committees that can address and resolve the issues they raise. This will lead to the production of a framework for developing a common vocabulary of definitions, best practices, protocols, and standards needed to support continued progress toward safer and more innovative automotive products.

The experts' meeting that gave rise to this publication demonstrated a great willingness on the part of the industry to define the terminology, procedures, and eventually the standards needed to enable automated electric vehicle integration technology and MaaS solutions to move ahead as quickly and efficiently as possible. SAE International has demonstrated its lead in this and closely related areas by the introduction of a basic reference work on active safety system sensors (SAE J3088™), the definitions for automated driving levels (SAE J3016™), the first vehicle cybersecurity framework (SAE J3061®), and numerous standards that establish the building blocks of ADAS, electrification, and intravehicular and extravehicular communications.

This SAE EDGE™ Research Report on "Unsettled Impacts of Integrating Automated Electric Vehicles into a Mobility-as-a-Service Ecosystem and Effects on Traditional Transportation and Ownership" identifies the following key topics for further pursuit, both through continued informal discussions among industry practitioners and through more formal working groups:

- Identifying the true cost of automated driving
- Developing AI and regulations for AV deployment
- Understanding the impact of electric vehicles on a MaaS ecosystem
- Determining where OEMs should compete and collaborate

Recommendations

The overall recommendations of this SAE EDGE™ Research Report can be summarized as follows:

1. Teleoperations, in lieu of safety drivers, could be used to maintain safety and reliability for low-speed automated shuttles in order to acclimate the public to MaaS platforms before extensive, high-speed applications are introduced.
2. Infrastructure like autonomous-vehicle operation zones that exclude driver-operated vehicles could help standardize driving scenarios, and establish a common language of understanding how vehicle systems should react to specific driving situations while true MaaS systems mature.
3. In order to test and validate highly automated systems, both closed and open testbeds are needed. Those testbeds must be compatible through the support of standardized scenarios and in full support of teleoperation through adequate V2X infrastructure.
4. Fleet operators using MaaS platforms need to plan the large-scale adoption of electrified vehicles and related infrastructure. This could be driven both by cost considerations as well as consumer demand, and will include technical training for operators and maintainers, as well as participation in the standards development process.
5. Public and private mobility service providers should collaborate to offer customers a complementary portfolio of transportation services that are cost effective and sustainable, while OEMs focus on supplying self-driving vehicles to service providers.

Abbreviations/Definitions

AAA - American Automobile Association

ADS - Automated Driving System. Intelligent computer control of vehicles based on sensor input, characteristic of vehicles at L3 and higher Automated Driving Levels. See the entry for this term below.

Automated Driving Levels (as defined by SAE J3016™)

L0 - No Driving Automation. A human driver performs the entire Dynamic Driving Task (DDT). Other vehicle systems may provide warnings or support, e.g., braking systems with automatic skid control.

L1 - Driver Assistance. The driving automation system is capable of executing either the longitudinal or the lateral vehicle motion control subtask (not both). It disengages immediately upon driver request. A human driver is present at all times to perform all control tasks not done by the automation system, to determine when it is appropriate to engage the automation system, to supervise all operations of the automation system, and to intervene as necessary to maintain safe vehicle operation. Example: automatic speed control ("cruise control") for highway driving.

L2 - Partial Driving Automation. The driving automation system is capable of executing both the lateral and the longitudinal vehicle motion control subtasks. It disengages immediately upon driver request. A human driver must be present and attentive at all times to perform control tasks not done by the automation system, to determine when it is appropriate to engage the automation system, to supervise all operations of the automation system, and to intervene as necessary to maintain safe vehicle operation. Total driver disengagement from human control mechanisms (hands off the wheel and feet off the pedals) is possible in narrowly limited driving scenarios, e.g., lane centering.

L3 - Conditional Driving Automation. A human driver must be present and attentive at all times to operate the vehicle outside the Operational Design Domain (ODD) of the Automated Driving System (ADS) and to take over control of the vehicle from the ADS within its ODD as necessary. The ADS can only be engaged in its ODD but can perform the entire DDT when engaged. It automatically disengages when it determines that either its ODD limits are about to be exceeded or it is experiencing a performance-relevant system failure, first, by issuing a timely request to intervene to the human driver and, second, by disengaging an appropriate time after issuing the request. It also disengages immediately upon driver request. The human driver is responsible for verifying the operational readiness of the vehicle, handling all DDT outside the ODD of the ADS, determining when the ADS can be engaged, and serving as the fallback-ready driver when the ADS is engaged, responsible for maintaining a minimal risk condition by actively intervening in ADS operation or assuming vehicle control on automatic ADS disengagement. An ADS ODD might include, for example, freeway driving or light traffic on open roads outside urban areas.

L4 - High Driving Automation. A human driver is responsible for verifying the operational readiness of the vehicle and determining whether to engage the ADS. The ADS will permit engagement only within its ODD. When the ADS is engaged, the human driver has no fallback role and may ride as a passenger. Once engaged, the ADS will transition to a minimal risk condition before it disengages 1) as the result of a performance-relevant system failure, 2) after an automatically issued request to intervene is ignored by a human driver, or 3) in response to a user request that it achieve a minimal risk condition. Optionally, a human driver may reassume DDT by requesting the ADS to disengage and waiting for it to turn over control of the vehicle once the ADS determines that conditions are appropriate; or, in a fallback role, following an ADS request to intervene; or, after the ADS has automatically disengaged, following achievement of a minimal risk condition.

F5 - Full Driving Automation. The ADS ODD expands to permit engagement under all driver-manageable on-road conditions. L5 vehicle operation, when the ADS is engaged, is identical to L4 vehicle operation.

AI - Artificial Intelligence. Computer systems that can exercise judgment and make decisions, e.g., recognize objects from visual inputs, understand spoken or written words and translate between languages, make optimal choices between alternative moves relative to achieving a goal (e.g., playing chess, navigating), etc.

AVSC - Automated Vehicle Safety Consortium

DDT - Dynamic Driving Task. Executing all the steering, acceleration, and braking functions needed to move a vehicle from one location to another safely.

IAMTS - International Alliance for Mobility Testing and Standardization

kWh - Kilowatt hour

MaaS - Mobility-as-a-Service

MIT - Massachusetts Institute of Technology

ODD - Operating Design (Driving) Domain, a specified scenario in which a self-driving vehicle can safely operate on its own, e.g., highway driving.

OEM - Original equipment manufacturer

Robotaxi - an SAE L4 or L5 autonomous vehicle operating as a ridesharing vehicle for a MaaS platform

US - United States

V2X - Vehicle-to-Everything communication. Among other possibilities, “X” can be another vehicle, a pedestrian, a roadside equipment, or the Internet.

Acknowledgments

Recognition should go first to all the participants, many of who also provided feedback on the draft version of this publication. Without their input and initiative, this SAE EDGE™ Research Report would not have been possible.

Dr. Robert Shorten - Imperial College London
Alexander Kraus - TÜV SÜD
Dan Baxter - Stantec
Ronald Grosse - Bertrandt
Dr. Reinhard Pfliegl - A3PS
Dr. Jim Martin - Clemson University
Mathew Yager - IOTA Foundation
Sean Kelley - Mannik & Smith Group
Hendrik Kramer - Technical University of Munich
Dr. Flavio Bonomi - Nebbiolo Technologies

The authors of this document together with the SAE Team responsible for its creation join in expressing our deepest appreciation to all the individuals mentioned above.

Dr. Joachim Taiber
International Transportation Innovation Center (ITIC), USA
Dr. Raivo Sell
Tallinn University of Technology, Estonia

Works Cited

1. Edmonds, E., "AAA: Ride-Hailing Twice the Cost of Car Ownership," AAA, https://newsroom.aaa.com/2018/08/ride-hailingdouble-cost-car-ownership/, accessed November 11, 2019.
2. Barter, P., "Cars Are Parked 95% of the Time. Let's Check!" Reinventing Parking, https://www.reinventingparking.org/2013/02/cars-areparked-95-of-time-lets-check.html, accessed November 11, 2019.
3. Baltic, T., Cappy, A., Hensley, R., and Pfaff, N., "How Sharing the Road Is Likely to Transform American Mobility," https://www.mckinsey.com/industries/automotiveand-assembly/our-insights/how-sharing-the-road-islikely-to-transform-american-mobility, accessed November 11, 2019.
4. Vehicle Technologies Office, "FOTW #1040, July 30, 2018: Average Vehicle Occupancy Remains Unchanged From 2009 to 2017," https://www.energy.gov/eere/vehicles/articles/fotw-1040-july-30-2018-average-vehicle-occupancyremains-unchanged-2009-2017, accessed November 11, 2019.
5. U.S. Department of Transportation, Federal Highway Administration, "Summary of Travel Trends: 2017 National Household Travel Survey," https://nhts.ornl.gov/assets/2017_nhts_summary_travel_trends.pdf, accessed November 11, 2019.
6. Flusche, D., "National Household Travel Survey -- Short Trips Analysis," News from The League, https://www.bikeleague.org/content/nationalhousehold-travel-survey-short-trips-analysis, accessed November 11, 2019.
7. Uber Technologies, Inc., Securities and Exchange Commission, https://www.sec.gov/Archives/edgar/data/1543151/000119312519103850/d647752ds1.htm, accessed November 11, 2019.
8. Del Rio, P. and Lank, O., "Study Shows Teens Delay Getting Driver's License," WSFB, https://www.wfsb.com/news/study-shows-teens-delaygetting-driver-s-license/article_472b3b5e-6dc8-11e9-b639-ef6043490806.html, accessed November 11, 2019.
9. Binette, J. and Vasold, K., "2018 Home and Community Preferences: A National Survey of Adults Age 18-Plus," AARP, https://www.aarp.org/research/topics/community/info-2018/2018-home-community-preference.html, accessed November 11, 2019.
10. Liyanage, S., Dia, H., Abduljabbar, R., and Bagloee, S.A., "Flexible Mobility On-Demand: An Environmental Scan," *Sustainability* 11(5):1262, 2019, doi:https://doi.org/10.3390/su11051262.
11. Davies, A., "The War to Remotely Control Self-Driving Cars Heats Up," Wired, https://www.wired.com/story/designated-driverteleoperations-self-driving-cars/, accessed November 11, 2019.
12. Hanley, S., "BNEF Shocker - Electric Cars Price Competitive In 2022 As Battery Costs Plummet," CleanTechnica, https://cleantechnica.com/2019/04/17/bnef-shockerelectric-cars-price-competitive-in-2020-as-batterycosts-plummet/, accessed November 11, 2019.
13. Myers, A., "How States Can Overcome the Looming Electric Vehicle Charging Infrastructure Gap: California," Forbes, https://www.forbes.com/sites/energyinnovation/2019/04/03/how-states-canovercome-the-looming-electric-vehicle-charginginfrastructure-gap/#38afb5d34ded, accessed November 11, 2019.
14. ChargePoint, "ChargePoint Express 250," https://www.chargepoint.com/products/commercial/cpe250/, accessed November 11, 2019.
15. Sivak, M. and Schoettle, B., "Relative Costs of Driving Electric and Gasoline Vehicles in the Individual U.S. States," University of Michigan, http://umich.edu/~umtriswt/PDF/SWT-2018-1.pdf, accessed November 11, 2019.
16. McMahon, J., "Electric Vehicles Cost Less than Half as Much to Drive," Forbes, https://www.forbes.com/sites/jeffmcmahon/2018/01/14/electric-vehicles-cost-less-than-half-as-much-todrive/#781c48983f97, accessed November 11, 2019.
17. Slowik, P., "Why Aren't Uber and Lyft All-Electric Already?" ICCT, https://theicct.org/blog/staff/why-arent-uber-and-lyftall-electric-already, accessed November 11, 2019.
18. Hannon, E., McKerracher, C., Orlandi, I., and Ramkumar, S., "An Integrated Perspective on the Future of Mobility," McKinsey, https://www.mckinsey.com/business-functions/sustainability/our-insights/an-integrated-perspective-on-the-future-of-mobility, accessed November 11, 2019.
19. Capgemini, "The Autonomous Car Report," https://www.capgemini.com/us-en/news/the-autonomous-car-report/, accessed November 11, 2019.
20. Narayanan, A., "Self-Driving Cars Run Into Reality - And Are Further Away than You Think," https://www.investors.com/news/self-driving-cars-hitdelays-driverless-cars-timeline/, accessed November 11, 2019.
21. Lambert, F., "Tesla unveils 'Robotaxi' Plan for Self-Driving Ride-Sharing Network Next Year," Electrek, https://electrek.co/2019/04/22/tesla-robotaxi-networkself-driving-fleet-ride-sharing-cars/, accessed November 11, 2019.
22. Automotive News, "MIT Study: Even with Robot Drivers, Ride-Hailing Still Expensive," https://www.autonews.com/mobility-report-newsletter/mit-study-even-robot-drivers-ride-hailing-stillexpensive, accessed November 11, 2019.
23. Nunes, A. and Hernandez, K.D., "Autonomous Vehicles and Public Health: High Cost or High

Opportunity Cost?" MIT, https://psyarxiv.com/6e94h, accessed November 11, 2019.

24. Etherington, D., TechCrunch, https://techcrunch.com/2017/11/20/uberorders-24000-volvo-xc90s-for-driverless-fleet/, accessed November 11, 2019.
25. Volvo, "Volvo Cars and Uber Present Production Vehicle Ready for Self-Driving," https://www.media.volvocars.com/global/en-gb/media/pressreleases/254697/volvo-cars-anduber-present-production-vehicle-ready-for-selfdriving, accessed November 11, 2019.
26. Esurance, "Free Time and Full Pockets: How Self-Driving Cars May Save You Money," https://www.esurance.com/insights/self-driving-carssave-money, accessed November 11, 2019.
27. Barry, K., "More American Drivers Want Electric Cars, AAA Survey Says," Consumer Reports, https://www.consumerreports.org/hybrids-evs/moreamerican-drivers-want-electric-cars-aaa-survey-says/, accessed November 11, 2019.
28. Robinson, R. and Tummalapalli, S.R., "Plugging into the Future: Electrifying the Global Automotive Industry," Deloitte, https://www2.deloitte.com/us/en/insights/industry/automotive/vehicle-electrification-global-automotiveindustry.html, accessed November 11, 2019.
29. Bigelow, P., "How Safe Do Self-Driving Vehicles Need to Be?" Automotive News, https://www.autonews.com/mobility-report/how-safedo-self-driving-vehicles-need-be, accessed November 11, 2019.
30. Soper, T., "BMW ReachNow Car-Sharing Service Shuts Down in Seattle and Portland Following Joint Venture Deal," GeekWire, https://www.geekwire.com/2019/bmw-reachnowshuts-car-sharing-service-seattle-portland-followingjoint-venture-deal/, accessed November 11, 2019.

Contact Information

EDGEresearch@sae.org.

The authors of this document together with the SAE Team responsible for its creation join in expressing our deepest appreciation to all of the individuals who contributed.

www.ingramcontent.com/pod-product-compliance
Lightning Source LLC
LaVergne TN
LVHW070941210926
848482LV00003B/276
* 9 7 8 1 4 6 8 6 0 1 1 0 7 *